The High Japanese Saving Rate: Causes and Prospects

Office of Trade and Investment Analysis
International Trade Administration
U.S. Department of Commerce

The High Japanese Saving Rate: Causes and Prospects

Office of Trade and Investment Analysis
International Trade Administration
U.S. Department of Commerce

Library of Congress Cataloging-in-Publication Data

Hijikata, Heidi.
  The High Japanese Saving Rate.   89-12154

    1. Saving and thrift--Japan. 2. Saving and investment--Japan.
I. Driscoll, Anne M. II. Fatheree, James W. III. Title.
HC465.S3H55   1988   332'.0415'0952   88-28546
ISBN 0-87186-343-X (pbk.)

First printing in bound-book form: 1988
First printed in the United States of America

PRICE: $3.50

COMMITTEE FOR ECONOMIC DEVELOPMENT
477 Madison Avenue, New York, N.Y. 10022 / (212) 688-2063
1700 K Street, N.W., Washington, D.C. 20006 / (202) 296-5860

CONTENTS

TEXT FIGURES

FOREWORD

This paper, <u>The High Japanese Saving Rate: Causes and Prospects</u>, was researched and written by Heidi Hijikata, Anne M. Driscoll, and James W. Fatheree of the U.S. Department of Commerce.

The Committee for Economic Development (CED) is publishing this paper as a public service to make it available to those working on issues relevant to U.S.-Japan economic relations. This paper is an analytical document and should not be construed as a statement of U.S. Department of Commerce or CED policy.

CED is an independent, nonprofit, and nonpartisan organization comprised of 250 corporate and academic leaders. Our members actively develop policy analysis and recommendations by blending their practical experience and background knowledge with the research capabilities of expert economists and social scientists.

CED, in collaboration with its Japanese counterpart organization, Keizai Doyukai (the Japan Association of Corporate Executives), has been studying various aspects of U.S.-Japanese economic relations for a number of years. The latest joint project of the two organizations is developing a semi-annual bulletin on U.S. and Japanese trade and economic developments.

Additional copies of the following paper can be ordered from CED in New York at (212) 688-2063. Questions on the content of the paper should be directed to Heidi Hijikata, Anne M. Driscoll, or James W. Fatheree at (202) 377-3124.

Robert C. Holland
President
Committee for Economic Development

SUMMARY AND KEY JUDGMENTS

The high saving rate among Japanese households has helped facilitate the emergence in recent years of Japan's huge trade and current account surpluses. Japanese saving behavior is thus relevant to the concerns of U.S. officials who deal with U.S. trade and other aspects of the United States' international economic and financial position.

Savings play a key role in determining a country's trade and current account balances because savings from income allow a country to use part of its production for domestic capital investment and exports. Furthermore, the savings that are used for capital outlays increase the size and improve the quality of a country's stock of plant and equipment, thus enhancing the competitiveness of its output in international markets.

The saving rate of Japanese households is high by international standards. OECD figures, for example, indicate that the gross household savings rate in Japan in the first half of the 1980s was over 16 percent, vs. about 10 percent in the United States.

Many factors explain why the Japanese household saving rate has been so high.

o   An unusually high proportion of the population has been in the working age category, a group whose saving rate tends to be higher than that of other age groups.

o   Cultural traditions, combined with relatively recent historical experiences--e.g., the hardship brought about by World War II--have developed strong inclinations to save.

o   A variety of current political and economic arrangements and circumstances have likewise promoted the high saving rate. These include, for example, the desire of most Japanese to own their own homes, combined with the high cost of land and limited access to credit; tax laws that encourage savings; and a long work week that discourages consumption.

The prospects appear reasonably good, however, that the Japanese household saving rate will decline in the future.

o   The proportion of the population in the older age brackets is on the rise, a development that is likely to reduce the saving rate.

o   Despite powerful pro-saving traditions, there is evidence that the population, particularly its younger elements, now takes a much more favorable view of consumption.

o   The Japanese government has taken several actions aimed at encouraging consumption at the expense of saving. These include new laws and regulations to increase the availability of consumer credit, reduce the work week, and change the tax system to weaken the incentive to save.

o   For the last two years, the government--largely in response to
     pressures triggered by the appreciation of the yen--has geared
     its overall economic policy to stimulating domestic demand.
     Though not explicitly aimed at reducing saving, Japan's
     expansionary measures have encouraged production for domestic
     uses, including consumption. Domestic demand did in fact grow
     rapidly in 1987--by 5 percent, according to official Japanese
     estimates--enabling GNP to grow by 3.7 percent, despite an
     almost 5 percent decline in real exports.

The anticipated decrease in the saving rate is likely to proceed
slowly, however. The changing age composition of the population
will have an effect only in the long run. Furthermore, neither
the depressant effect on savings of more favorable attitudes
toward consumption nor the reinforcing effect on these attitudes
of government efforts to inhibit savings and stimulate domestic
demand will occur overnight. In addition, formidable obstacles to
lower savings remain. For example, land and housing costs are
likely to remain high because of the preferential treatment the
government is likely to continue to accord agriculture.

INTRODUCTION

The high Japanese saving rate, because of its relation to the substantial current account surpluses Japan has run in recent years, is of considerable interest to U.S. officials concerned with U.S. trade and the U.S. international economic and financial position generally. Those surpluses grew from 0.7 percent of Japanese GNP in 1982 to 4.6 percent in 1986, before falling to about 3.7 percent in 1987. This paper summarizes the factors that have caused the Japanese saving rate to be so high in comparison with the rate elsewhere in the industrialized world, particularly the United States. It also assesses the prospects for a decline in the saving rate, examining factors--including policy actions by the Japanese government--that are likely to influence Japanese saving behavior in the future.

THE RELATIONSHIP BETWEEN THE JAPANESE CURRENT ACCOUNT SURPLUS AND THE HIGH SAVING RATE

By definition, savings that are not used for domestic investment must be exported. Japanese saving has consistently exceeded Japanese domestic investment needs, necessitating an outflow of capital from Japan. This outflow has slowed the appreciation of the yen, contributing to the continued international competitiveness of Japanese goods. At the same time, abundant Japanese saving has allowed for rapid expansion and improvement of the capital stock, also contributing to the competitiveness of Japan's goods.

U.S. AND JAPANESE SAVING RATES COMPARED

This section, for illustrative purposes, compares the national (or total) saving rate and its component parts for the United States and Japan for 1972-85. The components consist of saving by households, corporations, and government at all levels. Both gross and net saving rates are examined, to determine whether saving behavior over time and differences in saving behavior between the two countries are affected by the definition of saving used. Gross saving is defined as the portion of current income that is not paid in taxes or for consumption of goods and services. In financial terms, gross saving is the portion of income that flows into bank accounts, insurance policies, stocks and bonds and other financial assets. Borrowing and using accumulated saving for expenditures --dissaving--depresses the saving rate, while repayments increase it. Net saving is gross saving minus consumption (depreciation) of fixed capital.

Tables 1 and 2 Figures 1-6 lay out the basic data on the saving rates for the United States and Japan. Gross saving is presented as a proportion of gross national product, net saving as a proportion of net national product, which equals GNP minus consumption of fixed capital.

The major points brought out by these tables and figures are:

-Though the level of gross and net saving rates obviously differ, the pattern of movement over time for both variables for both the United States and Japan is very similar. Furthermore, the gap between Japanese and U.S. gross saving rates is not appreciably different from the disparity between Japanese and U.S. net saving rates.

-Both the U.S. and Japanese total saving rates were lower at the end compared with the beginning of the 1972-85 period.

-The Japanese national saving rate continued to exceed the U.S. national saving rate by a wide margin throughout the period. Indeed, the difference, though slightly lower in 1981-85 than in 1972-76, was higher in 1981-85 compared with 1977-80.

TABLE 1
U.S. AND JAPANESE SAVING RATES, 1972-1985
(Percent of GNP or NNP)a/

	Total b/ Japan	U.S.	Household c/ Japan	U.S.	Corporate Japan	U.S.	Governm Japan
1972-76							
Gross	35.3	18.5	19.5	10.3	10.8	8.1	5.0
net	25.4	8.8	18.1	8.0	2.2	2.4	5.2
1977-80							
Gross	31.7	19.9	18.6	10.3	10.4	9.0	2.7
Net	22.0	8.6	16.3	6.9	3.2	2.6	2.5
1981-85							
Gross	30.7	16.9	16.2	10.2	10.7	9.1	3.8
Net	20.0	4.4	13.6	6.7	2.7	2.0	3.7

a/ NNP is net national product, which is GNP minus consumption of fixed capital. Gross savings are calculated as a percentage of GNP. Net savi are calculated as a percentage of NNP. Gross savings are savings inclus of consumption of fixed capital. Net savings are savings net of consump of fixed capital.

b/ Totals may not equal sum of component parts because of rounding.

c/ Households include private unincorporated businesses.

Source: OECD, Department of Economics and Statistics, National Accounts 1973-85 and 1972-84.

TABLE 2
JAPANESE SAVING RATES MINUS
U.S. SAVING RATES, 1972-1985
(Percentage point differences)

	Total	Household	Corporate	Government
1972-76				
Gross	16.8	9.2	2.7	4.9
Net	16.6	10.1	-0.2	6.8
1977-80				
Gross	11.8	8.3	1.4	2.1
Net	13.4	9.4	0.6	3.4
1981-85				
Gross	13.8	6.0	1.6	6.2
net	15.6	6.9	0.7	8.0

Source: Derived from Table 1. See notes to Table 1.

- 4 -

Figure 1
Components of Japanese Gross Saving
as a Percentage of GNP

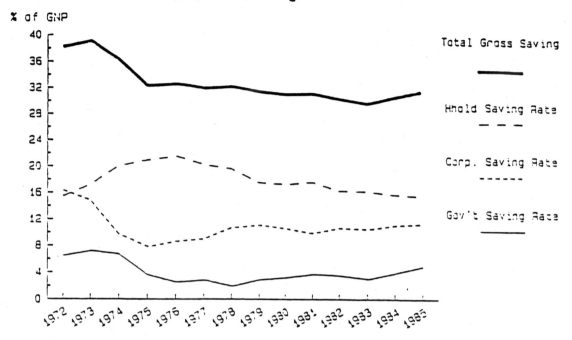

Figure 2
Components of U.S. Gross Saving
as a Percentage of GNP

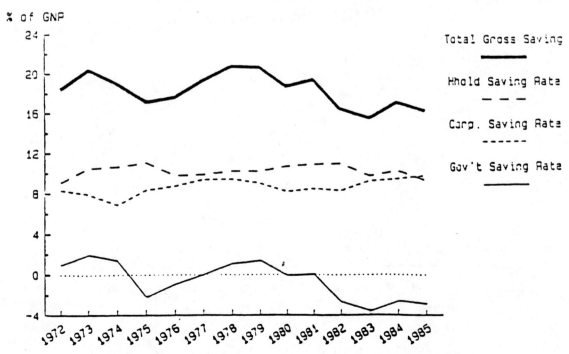

Figure 3
Components of Japanese Net Saving
as a Percentage of Net National Product

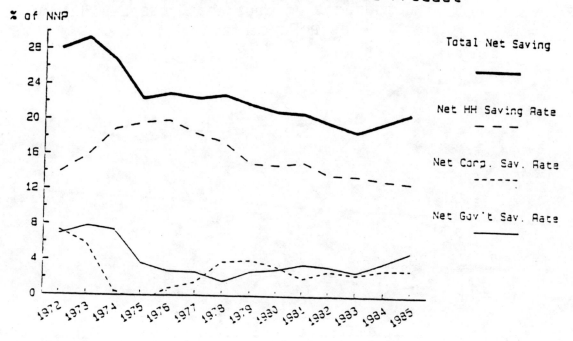

Figure 4
Components of U.S. Net Saving
as a Percentage of Net National Product

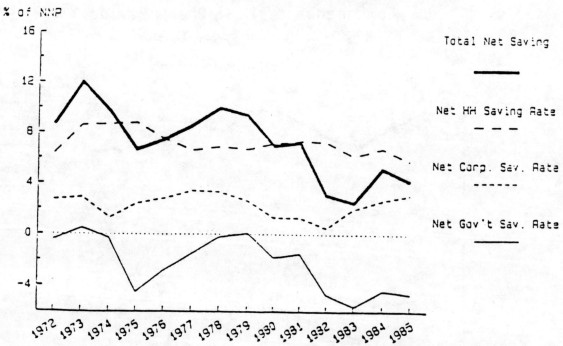

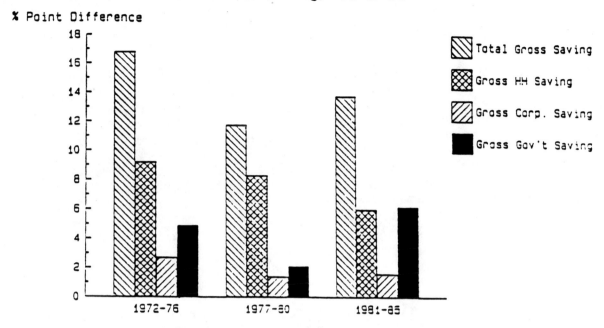

Figure 5
Japanese Saving Rates Minus U.S. Saving Rates
Gross Saving, 1972-85

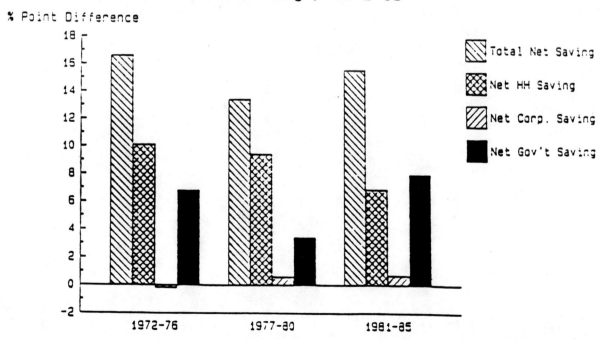

Figure 6
Japanese Saving Rates Minus U.S. Saving Rates
Net Savings, 1972-85

-The rise in the difference between the U.S. and Japanese national saving rates from 1977-80 to 1981-85 occurred in spite of a narrowing of the gap between the household saving rates of the two countries. The explanation is the steep increase in the disparity between the U.S. and Japanese government saving rates. The Japanese government saving rate rose between the two periods while that of the United States fell sharply and indeed, because of the federal budget deficit, was negative in 1981-85.

-In the 1972-80 period, the largest difference in U.S. and Japanese saving rates was in the household component. In 1981-85, however, the biggest disparity was in the government sector. Throughout the entire period, differences in the corporate saving rate of the two countries have been small.

Comparable U.S. and Japanese data for national saving rates and component parts are not available beyond 1985. However, comparable statistics, from the OECD, on household saving as a percentage of disposable personal income show a small rise for Japan and virtually no change for the United States from 1985 to 1986.

THE FOCUS OF THIS PAPER

The analysis in this paper will concentrate on the factors that determine the high household saving rate in Japan, primarily because households account for by far the largest share of total Japanese savings, on either a gross or net basis. In addition, except in recent years, the difference between U.S. and Japanese total saving rates has been mainly attributable to the large margin by which the Japanese household saving rate has surpassed the U.S. household saving rate. Moreover, this is likely to be true again in the future, in light of U.S. efforts to sharply reduce the federal budget deficit and the Japanese move toward a more expansionary fiscal policy.

FACTORS INFLUENCING THE HIGH JAPANESE HOUSEHOLD SAVING RATE

<u>Age Makeup of the Japanese Population</u>

The age makeup of the population has a substantial impact on a country's saving behavior. People nearing the end of their working years will typically save a relatively higher proportion of their income since salaries are at their peak and expenses are often starting to decrease. The young and elderly, because their incomes are normally lower, are likely to save little, or even dissave. Retired persons not only save little or nothing themselves, but they are also likely to depress the saving rates of their children, who often help to support them.

Compared to the United States as well as other industrial countries, a higher portion of the Japanese population is in its working years, especially relative to the portion of elderly.

-In 1984, almost 30 percent of the Japanese population was between the ages of 35 and 54 (compared to less than 23 percent of the U.S. population). Many of these people are approaching the end of their working careers--typically the time of greatest personal savings.

-But the share of the elderly in the population will be increasing. In 1984, about 10 percent of Japanese were over 64 compared with 12 percent of the U.S. population. By 2025, projections are that over 21 percent of the Japanese population will be over 64. In the United States, this segment of the population will comprise less than 20 percent of the population.

-As the Japanese population ages, a smaller percentage of the population will be in the highest saving bracket, while an increasing percentage of the population will be retired and more likely to be dissavers.

In sum, changes in the age structure of the Japanese population are expected to exert downward pressure on the saving rate. However, while the inhibiting effect on saving of this variable seems virtually certain, the impact is likely to occur only with a very substantial lag--perhaps a decade or more.

High Housing Costs

Although prices of land and houses in Japan are astronomical by the standards of most other industrial countries, about 62 percent of Japanese households own the dwelling in which they live. Relative to annual income, an average house in Tokyo is at least twice as expensive as a house in the United States. Thus, a homebuyer must devote a high percentage of his income toward the cost of purchasing his house. Accumulating savings to pay for the purchase of a house and making mortgage payments contribute significantly to the high saving rate.

Japan's steep real estate prices stem not only from a high population density and a shortage of usable land--much of the nation is mountainous--but also from the high degree of protection afforded to agriculture, which has aggravated the land shortage by limiting space available for housing construction. Although the Japanese government increasingly views more residential construction as an effective way to stimulate the economy and to improve the standard of living, the political strength of Japan's farmers appears to preclude extensive rezoning and removal of tax incentives for agricultural use of land. In the absence of reforms designed to make available more land for housing, prices will remain high. Since the purchase of a home remains a high priority for the Japanese, the high cost of housing will continue to exert strong upward pressure on the saving rate.

Retirement and Social Welfare Practices

In the past, Japan's social welfare and retirement provisions were very limited compared to most other industrial nations. This, too, has helped keep the Japanese household saving rate elevated, since individuals and families have saved more than they otherwise would have to protect themselves against emergencies and to prepare for old age. Since the early 1970s, however, Japan has upgraded its retirement programs. Although it still ranks low among the seven leading industrial nations in terms of social welfare spending--reflecting in part the low percentage of elderly in the population--rapid growth of Japanese spending on social welfare has brought Japan closer to international norms. Between FY 1973 and FY 1984, government social welfare spending increased at an average annual rate of 15.5 percent. In 1985, social security transfers in Japan were 11 percent of GDP, the same as in the United States. Furthermore, since the early 1970s, public pension benefits have been upgraded and now compare favorably with other countries.

The presumably depressant effect on savings of higher pension payments and welfare spending has been offset to some degree, however, by subsequent modifications. Because of the financial burden put on the system by an aging Japanese population, the government has recently undertaken reform of social security, reducing the maximum benefits allowable to a retiree. This will not lead to a decrease in average benefits received by retirees because most public pension recipients today have not been in the system long enough to receive the maximum amount. Although Japanese pension benefits still compare favorably with other industrial countries, the public perception of reduced benefits could lead individuals to save more.

Furthermore, in an effort to reduce the financial burden on the social security system, the government is considering raising the standard eligibility age for pensions from 60 to 65, thus encouraging workers to remain in the labor force for a longer period of time. The Japanese, who have the longest life expectancies in the industrial world, typically retire at 55 or 60.

Employment Practices--Japanese practices of long working hours and semiannual bonuses encourages saving.

The Work Week--The Japanese worked an average of 2,102 hours in 1986, compared with 1,924 in the United States and 1,600 in Europe, leaving Japanese workers with less time to spend their money than their European and American counterparts. In addition, the hours which Japanese businesses are open are often more limited than in the United States. This cuts further into time available for shopping or other consumption and thus has a positive impact on the saving rate.

In September 1987, the Japanese parliament enacted legislation decreasing the work week. Effective January 1988, the Japanese work week will be decreased by two hours to 46 hours. A further decrease to 44 hours is expected shortly thereafter. However, a significant portion of the Japanese working force will be exempted. Small companies--defined as those with less than 300 employees--which employ 85 percent of the Japanese labor force, will be given three years to comply with the new decree. Therefore, in the short run, this will have little effect on the saving rate.

Bonus system--Most workers in Japan receive semiannual bonuses as part of their compensation. Studies have shown that, in Japan, more is saved from bonus income than from regular wages. Workers generally base their budgeting decisions on their non-bonus income schedules, allowing bonus money to be used on "extras"--including savings. This system of compensation is not expected to change and therefore its effect on the saving rate should remain constant.

Disincentives for Consumption

<u>Limited consumer credit</u>--Until recently, Japan had little access to consumer credit. In the United States, on the other hand, consumer credit is widely available. This difference has been an important reason for the disparity between U.S. and Japanese household saving rates.

There is no necessary connection between lack of consumer credit and a high saving rate. As saving is defined as that part of income that is neither paid in taxes nor consumed, dissaving occurs whether an item is bought on credit or with accumulated savings. The net effect of both of these transactions on the saving rate is negative. Nevertheless, easy access to consumer credit may stimulate consumption at the expense of saving. Consumer credit allows consumption to take place in the present, rather than forcing it to be postponed to the future, thus allowing more purchases over a lifetime, and, more importantly, encouraging consumption by younger persons. In addition, consumer credit encourages consumption by allowing impulse buying. The lack of consumer credit discourages consumption and thus increases the saving rate. With specific purchases in mind, the Japanese are more likely to save large sums quickly--thus suppressing current consumption--since it will allow desired purchases to be made sooner. This likely leads to a higher saving rate than would result from loan repayments being made on a predetermined schedule.

Consumer credit in Japan--dominated by bank-affiliated credit card companies, credit sale firms, and small loan companies--is growing. Commercial banks are increasing their role, and it is hoped this increased competition will bring down the high interest rates which have discouraged borrowers. As of June, 1987, the combined balance of consumer loans extended by the twelve major Japanese commercial banks was up 70 percent over the previous year's level.

A portion of the increase in consumer credit is due to reduced interest rates, although installment credit and consumer lending rates have not fallen as far as other interest rates in Japan. Changing attitudes toward borrowing are credited with some of the increase. This trend toward increased borrowing is more pronounced among young Japanese, who appear more willing than their parents to go into debt.

The government has made several recent proposals aimed at increasing consumer credit:

-the removal of the 15 percent interest rate ceiling on money lent out by commercial banks. This would increase incentives to make money available for consumer credit and bring down the rates of moneylenders--who can charge up to 54.75 percent.

-the elimination of the prohibition against bank lending to consumer loan firms. This would lower the firms' lending costs and interest rates.

It is also being recommended that banks be allowed to offer their customers revolving credit, thus allowing credit card holders to pay back their debt in installments, rather than in one lump sum.

Increased availability of consumer credit is likely to encourage increased consumption. Expansion of consumer credit, furthermore, reflects a more positive Japanese view of consumption. It has been a change in attitudes--by the government, which is attempting to increase domestic demand, and the public, which has become more consumer oriented--that has brought about demands for increases in consumer credit availability. These two factors together--increased availability of consumer credit and changing attitudes--are likely to have a dampening effect on the Japanese saving rate. In the short run, this effect will not be great, as consumption patterns do not change overnight. But in the long-run, the effect could be significant.

Disincentives for Consumer Durable Purchases--The poor road system and "hutch-like" style of housing in Japan are disincentives for the purchases of consumer durables.

The Road System--The poor road system discourages purchase of automobiles. As part of their plan to stimulate domestic consumption, the government has initiated road-building and -improvement projects. However, upward spiraling land prices have increased the cost of these projects as a higher percentage of funds must go toward the purchase of land. This diminishes the scope of these programs because plans must be scaled back due to increased costs. The Japanese road system is likely to improve slowly at best and thus will generate little change in the Japanese saving rate.

The "Hutch-Like" Style of Housing--Small, expensive and inadequate housing discourages the purchase of consumer durables. In an attempt to encourage domestic consumption, the government is offering subsidized mortgage loans and tax incentives for new housing construction. However, increased costs of land dim the prospect for substantially increased demand for housing by the private sector. Thus a marked pickup in the demand for consumer durables is likewise unlikely.

The tax-exempt status of personal savings--The Japanese tax system
has long favored savers. Until recently, 58 to 70 percent of
personal financial savings were not taxed. However, as a result
of the recent tax reform bill, interest earned on small savings
accounts will be subject to tax, beginning in January, 1988.

The removal of the tax exemption from small savings accounts
takes away some of the financial incentives to save. However,
since much of the saving in Japan is targeted for specific
purchases or use--retirement, education, housing--it is doubtful
that this move will do much to drive down the saving rate.

Changing Family Size

A declining birth rate and a shift away from the traditional
extended family has caused the size of Japanese households to
decline in recent years. This could produce several effects on
the saving rate.

-Fewer children require less saving for education and
weddings, a factor which has often been cited as a reason for
the high Japanese saving rate.

-A smaller household has fewer expenses than a large
household. Thus, more income may be saved.

-Since fewer aged parents live with married children, the
expenses of the aged in the population have increased. This
presumably has a negative impact on the saving rate.

It is difficult to judge the net effect of these pressures on
the saving rate. It is quite possible that less saving for
education and weddings and increased dissavings by the elderly
would be offset by the increased saving resulting from reduced
consumption by families with fewer children. Therefore, it
appears unlikely that these factors will bring about great changes
in the Japanese saving rate.

Japanese Attitudes

The Japanese cultural heritage and Japan's perception of
itself as an economically vulnerable nation are often cited as
significant factors in the high Japanese saving rate.

-The Japanese cultural heritage, which encourages nonmaterial
values, is said to have slowed the rise in consumer
expectations and thus consumption. However, the Japanese
saving rate before World War II was significantly lower than
the post-war rate, suggesting that the inhibiting effect on
consumption of cultural heritage is limited or subject to
dilution over time.

-Although Japan is now a wealthy nation, the population in general does not consider itself well off. Feelings of economic vulnerability stem from the country's dependence on exports--Japan has few natural resources--and memories of economic hardship brought about by World War II. This is likely responsible for some precautionary saving, or saving for emergencies.

As memories of World War II fade, and the percentage of consumers growing up in an economically strong Japan increases, the perceived need for precautionary saving--and for policies which encourage saving--decreases. There are, in fact, clear signs of changing attitudes among the Japanese people and the government, as evidenced by a strong increase in consumer spending and some significant economic policy changes in 1987.

RECENT ECONOMIC DEVELOPMENTS AND POLICY MOVES IN JAPAN

Japanese economic policies for the last two years indicate a shift in national economic priorities. Reflecting pressures triggered by the steep appreciation of the yen since early 1985, Japan has sought to reduce its reliance on exports by stimulating domestic demand.

The Japanese economy, demonstrating considerable adaptability, has made visible progress in moving away from export dependence. Domestic demand, rising from 2 percent in 1983 to almost 5 percent in 1987, no longer lags behind GNP growth but paces it. Real GNP last year, despite a decline in real exports of 4.7 percent, rose by 3.7 percent. Forecasts for 1988 point to similar growth patterns.

Reflecting the rise in the yen and in domestic demand, the Japanese current account surplus, measured in yen, fell about 12 percent from 1986 to 1987. As a share of GNP, the surplus declined from 4.6 to 3.7 percent.

Stimulative fiscal and monetary actions have helped boost domestic demand. Recent measures include a June 1987 fiscal package that provides for almost 6 trillion yen ($45.6 billion) in public works spending for the fiscal year ending in March 1988; a September 1987 tax reform bill that reduced personal income taxes and cut tax privileges for interest income on small savings; and, in the monetary sphere, steps that increased liquidity and reduced interest rates following the October stock market crash.

It should be noted that Japanese economic growth last year was powered primarily by a sharp increase in investment. Outlays on plant and equipment rose by 7 percent, and there was also a sizable buildup of inventories. Lower interest rates also led to a sharp rise of 16 percent in private residential construction. Personal consumption increased last year at about the same rate as GNP, and, while figures are not yet available, it appears that the personal saving rate did not change much in 1987. Thus the surge in domestic demand and consequent decline in the current account surplus in 1987 in large measure stemmed not from more consumption and less saving but from a shift away from production for exports to production for domestic investment, with apparently little change in the saving rate.

However, the behavior of the economy and the policy measures taken last year suggest that the groundwork is being laid for a transition, over the long haul, to a much more consumer-oriented economy. With the less favorable outlook for exports, investment in plant and equipment is presumably being primarily directed at expanding capacity for production of goods for domestic consumption. Moreover, several of the recent expansionary policy actions--for example, the personal income tax cut and the move away from preferential tax treatment for interest income--seem clearly aimed at stimulating consumption.

Furthermore, there is mounting pressure from all levels of Japanese society to raise the standard of living. A notable example of this pressure was the 1986 report of the prestigious Maekawa Commission--so named after its chairman, a former head of the Bank of Japan. The report called for a restructuring of the Japanese economy, including a larger share of resources for consumption, to end the country's dependence on export-led growth. Though it gave the report a lukewarm reception, the Nakasone Government, and its successor the Takeshita Government, did, as noted, initiate policies designed to boost consumption. In addition, the Takeshita Government, through an Economic Council established in the wake of the Maekawa report, has begun drafting a new economic plan intended to improve the quality of life.

SUMMARY

The Japanese household saving rate has been edging downward since the 1970s but still remains high relative to that of the United States. The share of income Japanese households save is likely to continue to decrease for several reasons, including the rise in the years ahead of the proportion of the elderly in the population, policy measures by the Japanese government to encourage consumption at the expense of saving, and mounting pressure from the Japanese public to devote more resources to improving the standard of living. A declining saving rate should facilitate reduction of the Japanese current account surplus or, at least, of the ratio of the surplus to GNP.

Formidable obstacles to a reduced saving rate remain, however. An example is the continued preferential treatment accorded to agriculture, which elevates the cost of land and thus increases the savings required to buy a home. In addition, the forces working in favor of a diminished saving rate are not likely to have a speedy impact. The outlook thus appears to be for a gradual drop in the saving rate, with the extent of the drop difficult to determine.